黑虎爪單刀
Black Tiger Claw Single Saber

Dynamic Art of the Tiger

黑虎爪單刀
Black Tiger Claw Single Saber

Dynamic Art of the Tiger

Paul Koh

Kung Fu in a Minute Publishing

KUNG FU IN A MINUTE PUBLICATIONS

Copyright © 2023 Paul Koh

All rights reserved

如虎添翼
Yuet Fu Tiem Yik
Adding Wings to a Tiger

  Throughout the ages, the Chinese have been renowned for their phrases and sayings that can encapsulate an idea in four characters to convey a profound sense of meaning in a succint and short statement. The tiger is a symbol of something auspicious, great and powerful. This is a well-known theme throughout the Chinese culture as well as around the world, but this particular saying *Yuet Fu Tiem Yik* expresses taking that already powerful image and adding wings to it, making it that much more than it was before. This is the meaning behind this adage. This beautiful and modern interpretation of Chinese calligraphy was painted for me at the request of Grandmaster Tak Wah Eng to express the inspiration and energy imbued in this Black Tiger Single Saber set. It is as though the tiger has sprouted wings now and can fly.

Legendary Grandmaster Wai Hong of the Fu Jow Pai Tiger Claw System on the cover of Action Black Belt Magazine in 1974

Special Acknowledgments

Grandmaster Wai Hong's tiger claw & saber techniques passed down to him from his teacher, the late Great Grandmaster Wong Moon Toy, have inspired Master Tak Wah Eng to create the unique sets of saber fighting movements in this master text. Our gratitude to Grandmaster Wai Hong knows no bounds for teaching and giving us the ability to preserve the art for future generations. His skill and ability truly embody the visage and spirit of the tiger. With his guidance, we are able to bring forth the Dynamic Art of the Tiger.

The late Grandmaster Paul Eng demonstrating his single saber

SPECIAL ACKNOWLEDGMENTS

The background on this particular set is a unique composition inspired from the saber techniques of late Grandmaster Paul Eng and his teacher and mentor, our late Great Grandmaster Wong Moon Toy. The spirit and energy of the tiger truly reside within this particular set which is an exceptional and skillful display of classical Kung Fu weaponry.

DEDICATION

    The years it takes to learn, understand and master any aspect of the intricate art of Kung Fu is a daunting task to be taken up by many but reached by a very few. In my opinion, my teacher, Grandmaster Tak Wah Eng, is one of these special individuals that has persevered over many decades of hard work and training to achieve a true level of understanding and skill in this unique art form. Not only has he been able to reach a high level of personal skill in the various facets of the art of Kung Fu, but he has tirelessly worked to pass along his knowledge and passion to those of us that seek to grasp this elusive art. His ability to transcend all boundaries of different people's abilities to learn is a hallmark of a true teacher. His expertise in the weaponry of Kung Fu and its vast arsenal is truly astounding. Each weapon from the most simple and basic to the most esoteric come to life in his hands and weave a tapestry of beauty, grace and martial skill. He is able to bring out the genuine technique of each weapon as well as its innate beauty. The aspects of combat weaponry and exquisite technique are on full display, specifically in this wonderful single saber set inspired by the movements of Great Grandmaster Wong Moon Toy of the Fu Jow Pai Tiger Claw Kung Fu System and his Kung Fu legacy. I am truly indebted to my teacher for the endless bounty of knowledge he freely gives.

Preface

Back in the day when Kung Fu was becoming a household word but very few people truly understood what it was all about, there was no availability as there is today to find Kung Fu weaponry such as spears, straight swords, tiger forks, and the like. I remember when we first began to train in weaponry, there was no saber to be had so Sifu had to actually fashion one, literally, out of a piece of sheet metal. He handcrafted a wooden handle and a cup guard so we would have something to train with. That was the sole single saber in the whole school. Even though it was a homemade weapon, it was still highly prized by all and held in reverence because it was fashioned by his own hand and everyone had to take turns to play with it.

This anecdote reminds me strongly of how much we as Kung Fu practitioners should remember our beginnings and never grow too high and mighty in our own status and ability, always remembering to remain humble. Too many times, the younger generations don't understand the struggle and strife that the older generations had to go through to preserve the knowledge that they so freely partake in. Maintaining one's humility and respect towards whatever facet of the art of Kung Fu they learn is truly at the heart of what will help you to become a better martial artist and a better person all around.

True depth of knowledge and mastery must be always balanced with this reverence. Without this true and genuine gratitude towards your humble beginnings and the sacrifices of previous generations, you wouldn't be where you are today. In putting together this master text on the single saber, the phrase, "standing on the shoulders of giants" resounds in my mind. Without these forerunners that have come before us and set the groundwork, we truly would have nothing to speak of. The essence of the true warrior resides in his ability not only to be skillful with his martial technique, but maintain the proper balance, stemming from respect, honor and devotion to the masters of the past.

Table of Contents

1
Introduction

5
單刀的歷史
History of the Single Saber

11
虎魂
Soul of the Tiger

15
黑虎單刀要領
Essential Techniques of the Saber

31
黑虎爪單刀
Black Tiger Claw Single Saber

77
虎爪單刀對空手
Tiger Claw Saber vs. Empty Hand

105
單刀戰鬥技巧
Saber Techniques Against Staff

149
Conclusion

153
About the Author

Introduction

The great and massive arsenal of weaponry that is tied to Kung Fu is wholeheartedly misunderstood by the modern individual living in a relatively peaceful time. This great arsenal that is at the fingertips of Kung Fu masters has been handed down from generation to generation of martial artists to preserve lives and promote peace. Sounds extremely contrary to the fact. How can a weapon of death and destruction be able to promote peace? In order for us to understand this, we must understand the character 武 Wu or Mo, literally translating as martial. This character is comprised of various strokes that when studied and understood properly, do not promote war but rather strive to use the martial arts and their essence to stop or end the fighting, therefore striving to promote peace. The weapon itself must be viewed as simply a tool, no different than a hammer, a saw, a brush, a broom, and so on. The tool, placed in the proper hands of a well-trained martial artist with the correct moral and ethical fiber, can and will promote peace within society. Therefore, in the past, masters were highly selective in regards to who they would wish to teach and pass along their hard-earned skill to. This is the definition of what Kung Fu is—a skill set.

Even Kung Fu students and enthusiasts sometimes don't understand this. Many times, people want to learn a form without adopting the necessary skill set and foundation. In order to obtain the skill set of the saber, one's basics and empty hand training must be strong. Everything is predicated on how deep your basic techniques and foundation are. Your basic training bleeds through everything, and just learning this or any other weapons set is not going to make a difference unless the basics that are learned in the Tiger Claw System are in place.

When we practice Tiger Claw Kung Fu, everything is the tiger—every facet of your body, mind and spirit. We are not just executing separate gestures of hands and feet, stances and positions, but tapping into that true nature of a fighting tiger, which is all encompassing and then extends into the saber. Only then does it become a formidable weapon. As we will see in the following chapters, the attributes of the tiger and the Tiger Claw System will be seen in each and every action of the saber set. Every cut is there in your mind, big or small, circular, linear, high or low. Every move is representative of Tiger Claw Kung Fu. That's how tightly connected the weapon and the empty hand are in our system. There is no separation between the mindset of the tiger and the application of the weapon, but all the basic parameters have to be set first. Even if someone is to learn the saber set, without these basic ingredients, it won't look the same.

This is a concept of martial artistry which I think is lost in this generation. The term martial artist must denote the two seemingly opposite ideas, someone well-versed and skilled literally in combat but who can raise it to such a level in their execution that it has some artistic value to it. There's beauty involved. How is this not like a weapon itself? The weapon is a tool of destruction and death, but to behold a well-forged sword or saber, you marvel at its beauty, so much so that someone would hang it in an auspicious place. This is the dichotomy of learning Kung Fu. Make no mistake. When you practice these techniques, their inherent grace and beauty are only there because of the fostering of that deadly cut that is hiding behind every picturesque pose. That has to be wholly understood by the practitioner in order to gain the depth of understanding in learning this particular set and all other skill sets within Kung Fu. If you only practice the Kung Fu for its raw power to devastate an opponent without any philosophical or artistic principles behind it, you lose something in the process and vice versa. If you only practice it for its beauty then you wholeheartedly lose the soul of what this art form is created for and where it started.

Every movement that we do within the Tiger Claw System can be matched up or adjusted to possess any item in our hand which then becomes a weapon. But that notwithstanding, you must understand that every weapon and empty hand system has a particular skill set that one must learn, study and master. Too many individuals look at the single saber and because of its status as one of the four main weapons that are learned, view it as something simple, but as one spends time with the saber, the true skill set of Kung Fu and its proper utilization comes into play.

The skill required to wield any particular weapon from antiquity such as the saber requires an immense amount of time and energy. The expense of this work hopefully will resound within the individual, helping them to understand that their skill should be used as a deterrent and check for unscrupulous individuals within society rather than for wanton savagery. Be it in antiquity or today, this was and still is one of the underlying themes within training in the Chinese martial arts and its arsenal of weapons.

單刀的歷史
History of the Single Saber

A complete and accurate discussion of the art of Kung Fu cannot forego the thorough examination of weaponry and its training. Unlike what many may believe, the Asian martial arts and Kung Fu in particular do not deem their interests solely in hand-to-hand combat, but in fact, their practice is deeply interwoven with the usage of weaponry. So much so that many times, it is difficult to determine which was the genesis of the other. Did the usage of weapons influence that of the empty hand, or vice versa? So deep is this connection between weapon and empty hand training in the art of Kung Fu, to train solely in one aspect or the other is literally having only one hand or one leg.

One of the many reasons for this mandatory training in weaponry is indicated in some of the earliest dynastic records. Proficiency in weaponry was a necessity not only for survival but advancement through the ranks of the military and, of course, staying alive. It must be clearly understood that through the various dynasties of ancient China, men were required to serve in the military and to have a minimal of amount of rudimentary empty hand skills and of course weaponry. The history of Chinese weaponry is a profound one and has a direct impact on all aspects of the art of Kung Fu.

The most prolific and basic weapons are classified in four categories, the staff, the saber, the spear and the straight sword. These four elemental weapons comprise the basis for all other weapons utilized in Kung Fu systems, and uniquely, the saber stands as one of the most predominant found throughout. Many individuals who do not have training in the martial arts refer to any bladed weapon as a sword, but this is a grievous error. The straight sword is a double-edged bladed weapon that is used primarily for thrusting and slicing actions quite similar in some way or shape to modern day fencing, whereas the saber or broadsword, as it is called due to the flat, wider portion of the blade, is primarily for slashing, slicing, hacking and, of course, thrusting.

The saber has its origins in the most neolithic of history when men first began to shape stones into cutting implements that were used for survival, able to not only cut through tough hides, skins and meats of felled prey but also to be utilized for defense and protecting oneself and one's family. As ages passed and man began to learn the art of metallurgy, sabers and swords began to be crafted first in bronze then in iron, lending sturdiness, durability, the ability to hold an edge, and increased flexibility.

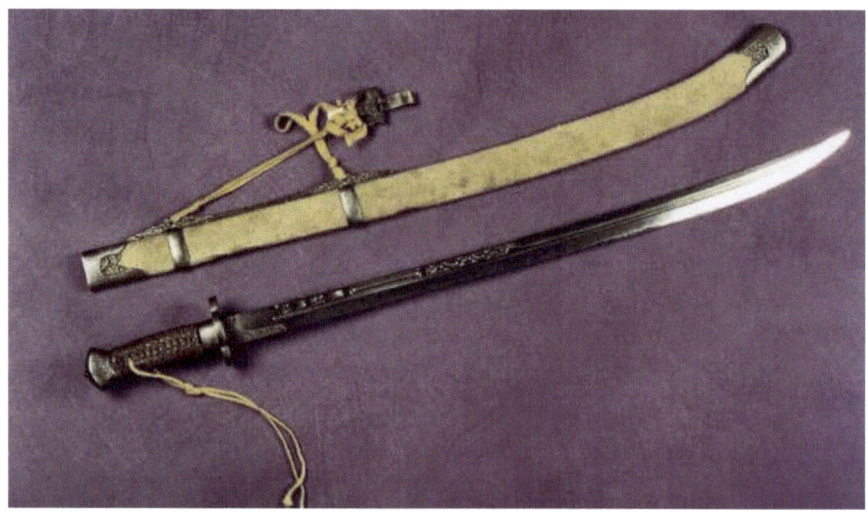

The saber/broadsword originally was a weapon coming from the north of China, slowly making its way down to the southern provinces. Over the centuries, the design of the Chinese saber has encompassed a wide variety of blades from thick to thin, from straighter edge and back to a more curved and broader design. The earliest recorded sabers that we know of had a straight back and became prevalent during the Shang Dynasty. This was the standard design for many centuries until the eventual invasion of China by the Mongols. The Mongolian warriors brought with them a curved back saber design that over time eclipsed the older, straight-backed sabers throughout the entire empire except upon its furthest fringes. This curved design is still prevalent even unto this day.

The saber can be seen as a side arm that would be primarily used in close-quarter combat after other weapons such as spears, halberds and the like are no longer usable. Each and every soldier, primarily foot soldiers, would be issued a saber and great training, dexterity and skill must be employed when utilizing the saber. Many times, we will see the single saber paired up with various shields, or in other instances, doubled up as a pair. The relevance and importance of the single saber must not be overlooked once the battle has been engaged and chariot fighters, armed cavalry and long-range weapons such as arrows and spears are spent. It would be up to the foot soldier in a one-on-one, eye-to-eye confrontation to win the day. Hence, the single saber and its practitioner were one of the most important aspects of ancient warfare.

The saber is a long bladed weapon with the back side being dull with a hilt and hand guard to protect the user. Perfectly balanced, the saber typifies the ferocity of a tiger in its slashing, hacking and thrusting ability. The single saber's structure allows for strong defensive and offensive maneuvers. The design of the blade gives a broad base for cutting along most of the surface of its edge. The blade of the single saber is long and can range from 25 to 38 inches. The blade is narrow but thick at its base closest to the handle and gradually widens as well as becoming razor thin getting closer to the top of the blade. The saber is sharp along most of the edge surface while the backside is almost completely dull to provide area for the practitioner to place his left hand and arm for specific techniques to the single saber, such as close range slicing and entrapping the opponent's weapon. The tip of the single saber is used for stabbing and thrusting at the opponent.

During the Ming and Ching dynastic periods, a common term for the single saber used by footsoldiers and martial artists alike was 腰刀 *yiu do*, waist-worn saber. Chinese imperial footsoldiers had particular regulations that their sabers must adhere to but they had freedom of choice of the style of the blade. During the 1700s, there were recorded to be approximately 19 variations of blade types. Unfortunately, there are no pictorial records of these varying blades. During the 19th century, the emergence of the 牛尾刀 *ngao mei do*, oxtail saber, became the most prevalent design used by most non-military individuals. It became the saber of choice for martial art heroes of the republic and rebels, slowly eclipsing all other designs and becoming the archetypal Chinese saber, still in use today.

The saber itself being one of the four basic weapons, was essential to the Kung Fu practitioner, basically because the saber itself, if viewed from a practical, modern standpoint, is a machete. What is the usage of such an implement? To hack and chop at vegetation, clearing forests. It was a tool, a farming implement, so the non-military Kung Fu practitioner could easily transition his thinking towards the application of this weapon. Even though many dynastic rulers outlawed military-grade weaponry for civilians, the saber became an indispensable tool, an everyday item that the Kung Fu practitioner would have access to. With the advent of the Manchurian Empire taking over China and systematically hunting down Ming rebels and supporters, the practice of weaponry became even more important to the Southern Shaolin rebels. Therefore, the emphasis on weaponry training took precedence, as guns were not common at this point in time.

The proficiency of the Kung Fu practitioner with weapons as well as in unarmed combat determined not only his ability to protect himself and his family, but was critical to his daily survival. The Kung Fu practitioner had to be well-versed in his weaponry training, because inevitably, aside from protecting himself, he would be faced with challenge matches from rival masters. Many times, these challenge matches would be designated as weapon vs. weapon. So, his ability to have command over a wide variety of weapons only raised the chances of him staying alive and surviving the encounter. It was integral, up even to the turn of the century, that Kung Fu masters have a good grasp of weaponry training.

Weaponry training is an invaluable asset that cannot be overlooked by the modern day Kung Fu practitioner. Specifically for the Tiger Claw Kung Fu student, the relationship of empty hand and weapon is deeply connected. Many schools take the perspective of reserving weaponry training as advanced material and do not teach weaponry to beginners while other experts insist that weaponry training taught from the beginning will help to develop speed, timing, body control, strength, grace and accuracy. The Tiger Claw System of Kung Fu will begin to teach weaponry early on in a student's journey to foster these qualities and underpin the inextricable connection that the techniques of the weapons derive and bolster those of the empty hand, and vice versa.

The saber being one of the four principal weapons of Kung Fu will be introduced to the tiger claw practitioner relatively early in his training regimen. The techniques of the tiger claw saber will teach and become an invaluable aid in a wide variety of specific and trademark techniques of the tiger, predominantly the stances and exclusive footwork as well as the strong empty hand chopping, clawing, ripping and tearing actions of the tiger that are hidden within the single saber set. The study and practice of the Tiger Claw Single Saber becomes a repository for the tiger claw practitioner to hone their skills in all these areas and further develop their understanding of the Dynamic Art of the Tiger.

虎魂
Soul of the Tiger

    All sentient beings strive to define and perceive what is life and the experience that they have with being alive. The definition of this is a philosophical conundrum that has baffled and puzzled mankind since his very beginning. In our struggle to define who and what we are, we have come up with terms such as spirit or soul, that intangible, unseen thing that animates all living beings. The soul, although unmeasurable in the terms of our modern science, was full well understood by the sages and philosophers of old. The soul or spirit of the tiger is at the core of everything that we, as practitioners of the Tiger Claw Kung Fu System, endeavor to do. Without this intangible connection to the spirit, each and every movement, regardless of empty hand or weapon, will fall flat and lifeless. The sheer, mere imitation of the visage of the tiger is far beyond the true scope of our practice. Imbued in the actions that we portray and execute in any of our sets, be they empty hand or weapon, we must be able to see this spirit or soul come alive. This is truly the case with this Tiger Claw Single Saber set. From the very beginning, we see the aspect of the tiger emanating from the form, traced in the silhouette of the body and shining in the eye. Each and every motion, slash and cut of the blade is permeated with the energetic spirit and voracious soul of the fighting tiger.

    Every stroke of the saber is drawing the image of the tiger. It's abstract, but the image is there nonetheless, even moreso than if we were to draw an exact picture of a tiger because it captures the essence of the animal, not just the silhouette. When painting the tiger, the artist has to be able to paint the bones, the sinew, the muscle of the tiger, not just the striped fur, so it is truly from within. When you first start to paint the image of the tiger, it falls short. You may draw everything to the exact specifications and not be able to duplicate what you see because what you see is not just the physical components but rather the energy. In time with proper practice, focus and guidance, the practitioner should begin not only to envision but feel that true essence come through into his movement.

    The saber itself has a clear and indelible connection to the tiger. The old masters always drew the analogy of the ferocity, speed, strength and energy of the saber to be aligned with that of the action and movement of the tiger. The actions of the saber imitate those of the tiger in its slashing, ripping and tearing, pouncing and spinning with agile steps. The saber, as the tiger, cuts down any and all adversaries.

    When playing the saber, one must adopt the mindset of the tiger. There is no quarter given with the saber. The whole concept of this weapon is to slaughter the opponent and hack and cut at any limb that protrudes outward. Therefore, its defensive movements and tactics are hidden in its offensive techniques. When learning and practicing the saber, the saber becomes the teacher and helps the practitioner reconcile mind and body to find that unique balance of hand, foot, mind and body coordination so every step, every cut, every thrust is as though it was done with ease, with no mind.

    The usage of the single saber will rely on the flexibility of the practitioner's body to be able to incorporate the weapon as such into his own movement. That is to say there is no separation whatsoever between the action of the saber practitioner and his tool or weapon. Aside from the movement of the footwork, stance and body, the practitioner utilizing the saber must view it as an extension of his own hand, therefore, complementing the movement issued forth from shoulder, elbow, wrist, extending into the saber itself. In this way, many saber techniques will take a direct cue from empty-hand movement. In our Tiger Claw Kung Fu System, it is well understood that we are able to adapt our movement to any weapon whatsoever in this way.

The saber is teaching you how to think and align your body, the movement and the blade together. It's a way of heightening one's awareness, which is a critical factor in the execution of any Kung Fu movement. In this way, the blade becomes the teacher and points out to you the mistakes or gaps that you may have in your movement. The blade teaches you where to be and where not to be and then you have to put this into your mind until every movement becomes effortless. You become like a whirling dervish, like a tiger coming out of the brush. Once he pounces on you, there's no stopping him. Once we begin that action of the cut that resides in the mind, there's no way to stop the blade. It is a constant, flowing action with a giant razor. Who can come close to the wild tiger and grasp him?

Within the movements of the saber, we see the shadow of the tiger released. The silhouette of the body with every cut and stroke shows the fierce attitude. The task of the practitioner is to find the tiger that resides within every stroke of the blade of the saber. This is incredibly important for practitioners of the Tiger Claw Kung Fu System to strike the unique balance between the weapon and our movement that is inherently imbued within.

The nature of the saber is to cut, and not just to cut them to bleed but to cut them to kill. You have to understand the nature of the saber and why is it so intertwined with the concept of the tiger. The claw of the tiger is just like the blade of the saber. If the claw of the tiger comes out, there is no way you are not getting cut to pieces. So we apply the same concept and idea to the saber. The tiger and the saber become synonymous through the practitioner's mind, heart, body and soul. When you see the person moving this set, you should see a tiger coming to life.

When practicing the Tiger Claw Saber, the practitioner becomes imbued with the spirit and nature of the tiger. The mind, the body, the blade all become one. The footwork is nimble, quick and ever-changing, allowing for no opportunity for the opponent to escape the cut of the blade. When we practice with the saber, our minds should be clear, clean and sharp just as the blade is. Understanding the pattern of the form, but not becoming entrapped within, the practitioner should be able to adjust and change each and every movement to fit the particular situation as needed. Once you become infused with the spirit of the tiger, you are no longer yourself. You become the blade and become the tiger.

黑虎單刀要領
Essential Techniques of the Saber

In the most rudimentary of senses, the practitioner learning the Black Tiger Single Saber must learn what end is up. Over the years, I have seen individuals employ saber techniques with straight swords and straight sword poses and techniques with sabers, clearly meaning that however they learned was wholeheartedly incorrect. The nature of each and every weapon is distinct and separate from the other. The old saying goes, they didn't know the difference between a hock and a hand saw.

There is a particular and specific way that the single saber is utilized, starting firstly with the grip, how the practitioner is holding it, the position of the body and stance, and of course, his mental demeanor. Even though every gun will shoot bullets, every gun or rifle will be held and shot in a different manner using different types of ammunition requiring a different mentality. So in this way, we must learn and appreciate the differences in all weapons. Therefore, we will be able to bring out the truer nature of the Black Tiger Claw Single Saber.

Looking at the structure of the single saber itself, its length, its breadth, the position of the handle and the hand guard, the dull back and the sharp edge, the practitioner must be aware and observe the different facets of the weapon to understand their usage and application. The weapon itself communicates to the player how it should be held and manipulated. Therefore, a great study must be done by anyone wishing to learn this or any weapon. We cannot take for granted or apply a general concept to the execution of the saber and its techniques.

The manipulation of the single saber goes far beyond grasping the handle and thrashing about as many may think. We have seen the image of children sword fighting back and forth with little wooden swords like many of us have done. This is not Kung Fu, as fun as it may be when you're a child. In fact, the incorporation of the entire body is what our focus should be upon, the saber itself being the last appendage in the execution of our movement.

The saber that we will employ in this form and throughout this book is termed the 牛尾刀 or *ngau mei do* or oxtail saber, which is favored by Chinese martial artists. Broader and wider at the end more than other sabers seen in earlier times, the oxtail saber allows for a wide variety of techniques. The various techniques that can be employed with the oxtail saber are hacking, chopping, thrusting, slashing, slicing, utilizing the entire length of the saber itself, as well as a unique technique of being able to support the back side of the saber with one's own palm to not only guide and deflect an oncoming blow, but use for slashing and cutting in an infighting, close quarter combat position. This type of fighting technique is *tong do*. It is a specialized type of movement that is featured within our Tiger Claw Single Saber set.

Everyone understands when they are looking at the saber that the edge is the most prominent if not the only aspect of the weapon, but this is not true. We have the edge, the tip, the back side and the handle. Many times, the handle is not considered a weapon, but if closely observed when held either in a ready position in

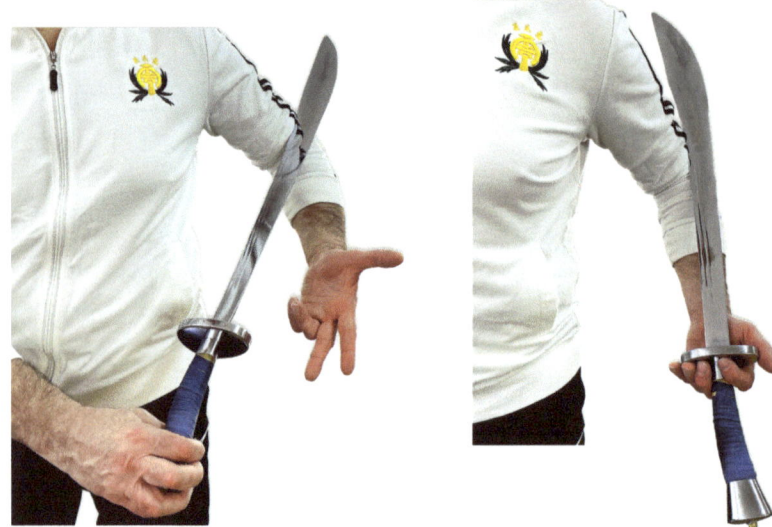

the left hand or in the predominant right hand, the pummel or handle of the blade is a formidable auxiliary weapon that can be looked upon synonymously with that of an elbow strike, a punch or a backfist.

Initially, these techniques may seem to be rudimentary in nature, yet this quality is essential to the single saber. Every technique is unadulterated and straightforward, in a sense put together succinctly for the true task at hand. Although there may seem to be flowery actions within its movement, one must not make the mistake to perceive them as such. The simplest movement of the knife is there for one sole purpose, and that is to cut. In truth, as said to me by my teacher, what is the use of the knife if it cannot cut?

When we begin to learn any individual saber technique, as mentioned before, we must understand grasping the saber handle itself. Utilizing the right hand, we will create a ring with the index finger and thumb as the major component of this new joint in the body with the three remaining fingers helping to control and guide the saber throughout its circular and linear techniques. The wrist must remain supple yet strong, the elbow flexible and the shoulder loose. The rest of the body also in this unique linkage must remain soft and supple, ready to be able to slash or strike out at a moment's notice. None of the movements of the saber should become stiff or rigid.

The limitations of the saber are the limitations of its practitioner. When the tiger claw student fully immerses himself in the study of the single saber and understands the nature of the tiger is residing within the saber set, all limitations can be removed and we will see the true visage and energy of the attacking tiger in the saber set.

虛步提刀
Hui Bo Tai Do
Tiger Claw Saber Ready Position

劈刀
Pek Do
Chopping Saber

過頭
Gor Tau
Encircling Head

削刀
Tong Do
Infighting Saber

挡刀
Dong Do
Blocking Saber

刀把
Do Bat
Saber Handle

黑虎爪單刀
BLACK TIGER SINGLE SABER

練刀之時。必須刀快眼快。心快步快。其來如風。其去如矢。身輕脚便洒脱自如全無凝滯。
自能得心應手。所向必克矣。

When practicing the saber set, you must be quick with your weapon, your gaze, your hands, and your steps. Come in like the wind, go out like an arrow. With your body light and your feet stepping with ease, there will automatically be no sluggishness in the movement. You will naturally be able to perform what is in your mind, and are sure to succeed.

— 金一明 Jin Yiming 單戒刀 Single Defense Saber, 1932

In Kung Fu, you must have the martial technique and at the same time, the individual must express his understanding. The physical technique is the structure which must be pristine and perfect. Then, you have to be able to modulate, move it and make it come alive. With the advent of modern day Wu Shu and its penchant for acrobatic and dance-like routines applied to the traditional weapons of Kung Fu, we have seen a downgrade of the genuine and authentic understanding and application of these true battlefield weapons. This is not a proper display of the true art of Kung Fu weaponry and its martial intent. Rather, it's become a prop for a gymnastic floor routine with martial accents.

Traditional Kung Fu practitioners must be wary of this and keep their mind focused on the martial content that is imbued by the nature of the weapon and stay true to that. Proper form, empty hand or weapon, without proper intention, is an exercise in futility. That should be clearly understood when learning Tiger Claw Kung Fu and any other facet such as this weapon is involved. What may appear to be a sword dance to an untrained individual is clearly not what we are doing. Awareness of oneself, the blade and the integration of the two into the form will help us to produce the proper end result. That end result not being only a well-executed and balanced set, but truly understanding how, when, where and why the cut is made.

With this understanding, any practitioner and especially one wishing to pursue studies in the Tiger Claw Kung Fu System must keep it foremost in their mind the true nature of any weapon is to kill. Displaying this intention and energy throughout every technique practiced in the saber set is one of the goals that we are striving for, remaining true to the nature of the weapon and the tiger at all times so we do not lose sight of what our practice is ultimately guiding us towards. The true essence of the Chinese martial arts is developing a greater understanding of oneself. We must hone the ability to utilize our martial technique while at the same time being able to remain calm yet resolute and not abuse the skill that we have learned.

The practice that one must do is a process of forging the body, mind and weapon into one unit. Many individuals are in a rush to learn the form thinking that they've digested the knowledge that's inside. You can rush to read the book but that doesn't mean you've read into every passage and extracted the meaning of the author's writing. This is my experience after learning many different saber sets. The essence of the movement is what we're after rather than the choreography of the form itself.

It is necessary to learn the sequence—this helps you establish the physcial presence of the movement, but we also must practice to capture the spirit and energy that must be injected into that movement. The essence without a coroproal body cannot be sustained, but what we are after is the understanding. This is in the heart and mind of the individual, which must bring the physical and spiritual together. The saber is teaching you that you are truly the weapon. It is a tool and a conduit for you to project your energy.

We can begin to understand the purpose of the saber techniques when we look at our execution from the outside. The saber and the form are essentially a magnifying glass. They serve to magnify the rhythm and timing so we see where the individual is in proper alignment and misalignment because you have this inanimnate object that must be assimilated into the human body. If it's done well, the rhythm and timing are unbroken. If performed improperly, then the form becomes mechanical and clumsy. If the practitioner is observant enough, he can correct his rhythm and timing based on the feedback from the saber because there really should be no break in connection from the hand to the saber. The player has to spend time to observe themselves and be aware of their rhythm, timing, step, and the way they execute any particular movement, so there is a lot of introspective correction that has to take place. The teacher resides in the saber and its movement, but is the student willing to listen and look to make those corrections? That is really at the heart of the issue for learning this saber, heightening the awareness of the practitioner of himself so that awareness can flow into the saber or whatever weapon he's performing.

Circles & Lines

When viewing this weapon set, the practitioner should take a step back to see the physical structure of each and every technique and how they are put together and related to one another in the sense of linear and circular movements. Circles and lines, these basic components of geometry, are seen throughout the Black Tiger Single Saber set. A deep understanding of how these circles and lines are configured in the slashing, cutting, stabbing actions of the saber is necessary in order for the tiger claw practitioner to not only execute and perfect his movement but to better understand the cut, which is at the core of this weapon's execution.

Circular movements will be executed in large and small variations. The large, circular movements come from the base of the shoulder and can rise upward in a vertical circle or downward as well as cutting in a horizontal circle parallel to the ground. Fostering these circular motions will be the elbow and the wrist. The wrist comes into play for smaller, faster circular techniques that can be applied in a flowering way, creating a figure eight cutting pattern that can rise or fall as needed. We now see with these circular patterns of vertical, horizontal and diagonal that we are adhering to the nature of the 八卦 *bhat gwa,* and this eight diagram formation is engulfing the entire body of the practitioner. Complete suppleness and relaxation of the entire body as well as mind is required to employ these circular cutting patterns, thereby making the saber practitioner and his weapon inseparable.

Straight lines or linear techniques may seem many times to be an outgrowth of the circular and vice versa. Linear techniques can be readily observed in the cutting, chopping and stabbing actions of the saber. Thrusting out with the tip, slicing and hacking are all techniques that will draw a line with the length of the saber's blade. These cutting techniques also conform to a vertical up or down cut, as well as horizontal and diagonal with the added feature of stabbing and thrusting directly out from the practitioner, either in a high, middle or low trajectory. It must be understood that we will not make a harsh division between circular and linear techniques of the single saber yet each reside within the other and foster each other in their execution. A linear thrust may be followed up quickly by a circular slash. A circular wrapping cut can be followed up immediately by a linear chopping strike. In this way, we see the circular and linear techniques of the Black Tiger Single Saber are inseparable.

Don't Forget the Left Hand

Aside from the saber being the major, obvious player in the execution of these techniques, we must understand the rest of the body will come into play, and the left hand will become the balancing component to the saber. Again, we see the emergence of the harmonious balance between 陰陽 *yum yeung* or *yin* and *yang*, in this instance, right and left hand, the right hand being the *yeung* element—forceful, fierce, strong and fast—the left hand being the *yum*—softer, less apparent, yet supporting every action of the right hand. The left hand acts as a balancing point when deflecting, blocking, wrapping around the body, helping to open, obstruct, as well as in many instances, grasping the opponent or his weapon so the right hand with saber can execute a final cutting blow. In this way, the left hand is just as integral to the execution of the saber form as the right hand itself. *Yum* does not exist without *yeung* and vice versa. Once we have digested this concept, we can apply it throughout the entire saber set and every movement therein.

Mind Your Mind

When applying the action and movement of the saber, the entire body is intertwined with the movement, specifically in the cutting and slashing. The entire arm, now accentuated by the length of the saber, will be used, from the shoulder, elbow, wrist, hand, extending all the way into the blade itself. In that way, the blade has now been assimilated into the body of the practitioner and there is no separation whatsoever. This is no different than the skill level of a surgeon with his scalpel. The mind of the practitioner leads directly into the blade, and the blade becomes one with him.

Learning this particular saber set or any weapon is a huge accomplishment, but the work begins at the threshold of finishing the set. Now that the physical parameters have been set before us, we need to begin to understand the internal processes that must be embraced by the practitioner to fully flesh out the true understanding hidden within this unique weapon. One may possess the blade. One may learn the steps and actions of the form. One may even practice fervently on a daily basis. Yet when one is lacking the mind, the movement and action of the Black Tiger Single Saber will never come alive.

The live tiger hidden within the blade is there to make the cut. The cut is in the mind of the practitioner. We can use the analogy of a paper cut. All of us at one point in time have given ourselves a paper cut, and you say to yourself, well, how did the paper cut me? It's just a piece of paper. It's thin, seemingly fragile, easily torn, crumpled up and destroyed but yet, with the slightest of movements and no pressure or force at all, it cuts you. This is a very clear example of how the mind must perceive the cutting action of the Black Tiger Single Saber. A comment was made to me by a senior student when we were practicing this particular form and I demonstrated to him the section of the set that we were going over at that time. His remark to me was, "Sifu, it seems like you're really putting all your energy and power into it, but yet it's so swift and light, almost effortless, while I'm trying so hard." Therein lies a secret, a visible secret that's always been in front of your face, but yet we refuse to recognize it. Sometimes, trying too hard is the complete opposite of what we need to do.

The blade is the mind. The mind is in the blade. If the mind is sharp, the blade is sharp. If the mind is dull, the blade is dull. How we draw the saber and enact it with our mind in all of the techniques that we learn produces the end result. Did we make the cut or fail to cut? In this way, we must strive to understand what the cut is. The cut is just a natural stroke of the hand, be it with saber, sword, claw or brush. It is the most natural and simplest of actions, but because we are so intent on trying hard to get it, we miss it completely and end up muscling through the movement where no such effort was truly required. A tiger's every action is flowing, embodied with fierce kinetic energy, never stagnant nor mechanical. The saber also should strive to embody this idea. Thereby, we find the true balance of harmony within the Black Tiger Single Saber set and the Dynamic Art of the Tiger.

The photographs displayed here depicting the *Hark Fu Jow Dan Do* have been shot specifically to show the entire scope and range of the movements and techniques found in this set. They are shot in a particular way to depict the positions as well as the transitions from one to another to give a fuller, more well-rounded view and understanding of how the individual saber techniques will be executed. These series of photos are laid out in a step-by-step manner. The dynamic photography and layout is there to provide the viewer and those truly interested in learning the set with an understanding that the form itself is alive. It has moments of completion for a particular technique, but in between these positions, we find the transitory movements that are actually providing instruction on how the saber is executing its individual strikes or cutting techniques. A pose itself will not provide enough feedback for the reader to truly understand the execution of the movment. We sincerely hope that this display helps those that are studying from this book who have either learned or wish to learn this set to have a greater resource and foster their understanding of the movements within the saber set.

黑虎爪單刀

BLACK TIGER CLAW
SINGLE SABER

黑虎爪單刀 從頭
Black Tiger Claw Single Saber Salutation

1. Feet together at the heels. Raise up both hands.

2. Step forward with right foot into left tiger stance, bringing both hands forward to heart level, right hand in knife hand position, left hand holding saber.

3A–C. Draw right hand back in sword finger position, protecting head. Immediately, raise left hand with saber to block left side. Follow up directly with right uppercut.

4A–C. Maintain right hand in uppercut position. Circle left foot back into right cross stance. Circle left foot forward, hooking in. Immediately execute left front kick.

5. Following up from front kick, step into left bow, bringing saber to left side in line with stance, striking out with right single tiger claw, shoulder level.

6. Shift body to back right corner in right tiger stance, executing saber block with handle and right low claw hand.

7A–C. Step with left foot into left cross stance facing front left corner, rotating saber downward as right hand rises. Step up with right foot into right cross stance, executing inside tiger claw, drawing right hand to waist, saber to eye level. Immediately, step forward into left front stance, drawing saber hand into body with blade facing otuward, executing tiger claw strike with right.

8. Execute inside tiger claw block, stepping forward into left tiger stance facing front right corner, ending in single tiger claw strike.
9. Step over with right foot into reverse cross stance, executing tiger claw strike facing directly front.

10A–C. Execute left sweeping kick and right inside mirror hand, stepping towards back. Execute right sweeping kick with left inside saber blocking hand, stepping towards back. Place right foot down. Turn body around to face front in reverse bow stance position.

黑虎爪單刀 一路
Black Tiger Claw Single Saber Section 1

11. Grasping saber with right hand, perform slashing cut to back. Spin a full 360 degrees into stabbing stance with extended low cut.

12. Step forward with left foot, executing cut with ridge of saber to left side.

13. Without stopping, continue spin while wrapping around head and neck, completing second 360 spin and executing cross body cut to left side.

14. Execute three consecutive upward slashing cuts, walking towards front into right tiger stance, saber resting on shoulder.

14

15A–E. Step back with left foot, executing low slashing cut. Immediately, bring saber up to wrap around head. Roll saber downward to make circular cut into right front stance with thrusting stabbing strike facing back.

15C

15D

15E

16A–C. Roll saber with wrist to right side, stepping to back into cross stance. Rotate entire body to cut over to back. Execute three upward slashing strikes to step forward into right bow stance with final upward slash to front.

16C

17A

17A–B. From previous upward slash, turn to back, executing two wide arching circular downward cuts stepping forward into overhand stab.
18A–B. Continuing from downward stab, execute four inside saber deflections from left to right, culminating in high knife hand saber block.

17B

18A

19A–B. From high saber block, step forward, executing horizontal circular cut, rotating backward two times, stepping back into left cross stance with diagonal cut, saber wrapping around body.
20. From left cross stance position, execute high swinging heel kick.

21A–F. Following heel kick, step back with left foot, extending left hand to face front and begin three downward chopping rotations in right tiger stance. Reverse wrist to place dull edge of blade on waist, turning torso to execute parallel waist cut in reverse bow stance. Follow up with left foot stepping behind, executing large circular downward slashing cut. Step back with left foot, turning 360 degrees to commence second flower cutting sequence. This sequence occurs three times consecutively.

21A

21B

21C

21D

21E

21F

22A–E. Following last flower rotation, turn body and execute horizontal cut wrapping around head two times to culminate in left cross stance facing back, simultaneously placing saber in left hand.

22A

22B

23. Execute high front kick, slapping right hand to instep of foot.

24. Turn body around to face directly front, drawing saber back to left shoulder and executing right tiger claw strike in reverse bow stance.

黑虎爪單刀 二路
Black Tiger Claw Single Saber Section 2

25A–C. From tiger claw strike, turn torso to left and right. Then, step back into right cross stance with saber behind back and high right knife hand, facing back.

26. Turn body around, grasping saber to complete rotation facing back, executing deflecting block, panning face from left to right, culminating in stabbing strike in right bow facing directly back.

27A–E. Draw saber backward while stepping back with right foot to place palm on back of blade. Immediately step up with left foot to slash upward. Bring saber over to face front, cutting downward. Continue cut, turning body to face back, slashing upward. Step behind with left foot into low cross stance, executing low cut to front with saber.

28A–B. From cross stance, step forward with left foot, wrapping saber around head to turn entire body to face right side, following up with diagonal slashing cut.

29A–D. Raise right foot. Execute low diagonal cut, bringing saber to shoulder. Step down. Wrap around head and cut facing right. Turn swiftly round, executing diagonal cut to left, wrapping saber around body.

29D

30A–E. From last diagonal cut, bring right foot forward into tiger stance and execute large circular cutting strike from shoulder with both hands. Bring blade and left hand forward into body, executing several smaller rotating cuts with wrist, crisscrossing in front of torso.

30E

31A

31A–B. Following last criss-cross cut, execute deflecting motion with saber upward, following up with upper slashing cut in right bow facing directly left.

32A–B. Step forward with left foot into cross stance. Execute right kick to dull side of blade.

33. Step back into low right cross stance with low cut facing left side.
34. Turn body to face right side. Execute upward slashing cut with left rising tiger claw and sweeping kick.
35. Execute skipping jump into low right back stance with slashing cut and tiger claw.

36A–C. Draw left foot and hand into body, executing right inside crescent kick to front.

37A–B. Directly following crescent kick, wrap saber around head and turn to face back, executing diagonal saber cut.

38. Execute low slashing cut to left, simultaneously turning around 360 degrees, bringing blade to side of body.

39. Draw saber from left side to right side, bracing with left hand for high block in crane stance, blade facing to front.

40. Step down and backward, executing low block in cross stance to face back.

黑虎爪單刀 三路
Black Tiger Claw Single Saber Section 3

41A–B. From low blocking position, wrap blade around head, executing horizontal cut, followed up with left hook hand position in left bow stance facing back.

42. Maintaining hook hand and blade position, execute high swinging heel kick with right leg.
43. Bring saber around in counter clockwise motion to protect face. Step backward into right cross stance, executing level saber cut with left claw to left side.
44. Step up with high brushing block with saber into upward slash in right bow stance facing back.

45A–B. With right foot, step forward. Turn torso backward, executing overhand cut to back. Immediately, continue cut and step forward into right bow, finishing up with upward slashing cut to front.

46A–C. Shift weight back to left leg, assuming right tiger stance, executing three reverse saber deflections.

46A

46B

46C

47A–C. Following last deflection, rotate blade to place dull edge on waist, turning torso to execute horizontal waist cut. Immediately wrap around head, spinning body 360 degrees to face forward in left tiger stance with high chopping strike.

48. From high chop, roll wrist counterclockwise to place saber in left hand, handle between index and middle finger. Step into left bow, executing single right tiger claw strike.

49. Step backward into right cross stance, executing tiger claw strike directly right.

黑虎爪單刀 結論
BLACK TIGER CLAW SINGLE SABER CLOSING

50. Step out into left bow facing left, executing low tiger claw strike to groin.
51. Immediately, execute overhand tiger claw strike.
52. Draw left foot back into toe stance, saber in line with body and right claw to temple level.

53. Step out with left leg, executing strike with pummel to back. **54.** Continue immediately, stepping towards back with right foot and execute downward diagonal tiger claw ripping strike to right side. **55A–B.** Shift body and execute backhand strike with pummel. Follow up immediately with single tiger claw strike facing right.

56. Turn torso back and then front to face forward, high right knife hand, saber parallel to floor in reverse bow stance.
57. Draw body up into left tiger stance, crossing torso with saber and tiger claw, bringing both to either side.
58. Sink bridge hand on both left and right sides.
59. Draw saber and right claw inward to heart level, executing final closing tiger claw salutation.

Tiger Claw Saber vs. Empty Hand
虎爪單刀對空手

Matching and fighting sets are at the core of many, if not most, Chinese martial art systems, endeavoring to teach and imbue their students with the proper understanding and principles derived from their techniques to help them develop and understand how to apply them for the benefit of self-defense. Unfortunately, many times these matching sets take on an overly dramatic and theatrical aspect that undermines their true practicality. This is not the case with this particular matching set. The genesis of this and other practical matching sets will be at its core functional techniques that can be applied either against bare handed attacks or attacks stemming from weaponry.

In displaying this particular saber vs. empty hand matching set, I wish to take the position of devil's advocate. We need to be quite clear and truthful with ourselves to understand that it will be extremely dangerous and difficult if not almost impossible to defend against an opponent wielding a machete-sized knife completely bare handed. That is not to say that it is completely unfeasible. With this understanding in mind, the rational person may come forward and ask, "Then what is the purpose of having such a matching set and learning the empty-handed side if the danger of being cut into pieces is so prevalent?" And they would be right to ask such a question. In defense of the reasoning and logic of why one should invest time in learning and practicing such a matching set, we must understand what are the benefits derived from learning such a set.

The heightened danger presented to the empty-hand player is truly what we are looking for. This will benefit both partners, serving to increase their awareness of self, surroundings and the eminent danger of being cut. Conversely, the saber player can begin to understand and visualize how his techniques can and will be applied. Intrinsically, the practice of this matching set filters down invaluable benefits to both parties. We can view this matching set almost in a *yin* and *yang* scenario where the wielder of the saber is

in actuality performing empty hand movements and the empty-handed partner is truly executing movements that can be suited for weaponry as well as bare-handed fighting. Timing, speed, agility, understanding of distance will all be heightened and honed to a higher degree. All the senses of both partners over time will be brought to an instinctual and reactive level that we hope will take root in them and proliferate throughout their entire practice regardless of what facet of the Dynamic Art of the Tiger they

delve into. When the two partners practice this matching set, they will push and prompt one another and eventually develop and transform into one another, again bringing about this *yin* and *yang* quality where one aspect of each resides in the other, thereby creating a harmonious balance on the edge of a knife.

The matching set will force each individual to learn to move swiftly without obstruction and break all confines of stagnant poses that may have been misconstrued from learning empty hand forms and the like. One wrong move and either one of the participants can incur injury. Therefore, when practicing the saber vs. empty hand set, great care must be taken on either side to ensure this does not happen. Over time, the set will be played faster, more fervently, with more energy, until it takes on the shape of realism. It is suggested when practicing this and any matching set that smaller segments should be taken out of the context of the greater whole and practiced individually to capture the realistic sense of a self-defense scenario, be it either side. Step by step, both practitioners will be able not only to learn and master the individual techniques but be able to understand and produce free form techniques at will, which is the true meaning of learning any type of two-man fighting set such as this.

The empty hand mirrors many of the actions of the saber and vice versa. In this way, they are akin to a mirror image of each other but not exactly the same. The actions of the weapon and/or empty hand techniques found within the Tiger Claw System reside in one another and can be seen to grow from one to the other and extend to each other. This can be seen in the complementary matching set of saber vs. empty hand. In this way, we truly understand it as a cohesive system where there is no separation of principles of the tiger claw technique, but rather, it is the hingepin or underlying theme that runs throughout all the skill sets within the Dynamic Art of the Tiger. When we observe two well-skilled and evenly matched partners, it is as though we are witnessing two fierce tigers in a battle to survive.

Sifu Koh practicing Tiger Claw Saber vs. Empty Hand with Grandmaster Tak Wah Eng, mid 2000s.

虎爪單刀對空手

TIGER CLAW SABER
VS. EMPTY HAND

虎爪單刀對空手 從頭
Tiger Claw Saber vs. Empty Hand
Opening Salutation

Both partners step out into left bow facing one another, executing descending left tiger claw block with right tiger claw strike.

Both partners shift to their right, executing high and low tiger claw deflection.

Both partners step over to their left, executing high and low tiger claw deflection in cross stance.

Both partners step over into right cross stance with left high tiger claw tear.

Both partners step forward into left bow, following up with double tiger claw strike.

Both partners step to their right, executing single tiger claw strike in left tiger stance.

Both partners step away from one another into right reverse cross stance, executing tiger claw strike.

Both partners step backward to face each other in low back stance with high knife hand position.

Both partners rise up into left tiger stance, executing right and left downward facing claws.

Remaining in stance, both partners sink bridge hands.

Both partners draw hands inward to present tiger claw salutation.

Both partners assume tiger claw ready position.

虎爪單刀對空手 一路
Tiger Claw Saber vs. Empty Hand Part 1

Saber steps forward, executing right diagonal chop. **Empty Hand** shifts to his left, executing double tiger claw palm block to inside forearm of saber attack.

Saber applies tiger claw bridge hand from underneath to shear off double palm block.

Saber immediately comes over to execute diagonal chop on other side. **Empty Hand** shifts to his right side, executing double tiger palm block to outside of saber attack.

Empty Hand executes double palm push to saber attack, preparing for foot sweep.

Empty Hand follows up with right foot sweep as **Saber** raises right leg to evade.

Continuing with sweep, **Empty Hand** spins 360 degrees. As **Saber** follows through with evasion, **Saber** also spins 360 degrees.

Both partners finish rotation. **Saber** executes high chop to head as **Empty Hand** blocks with tiger claw high X block.

Empty Hand uses left hand to grasp wrist of **Saber's** attack, diverting saber down and away to his right.

Saber uses this motion to attempt slashing cut to **Empty Hand's** right leg. **Empty Hand** raises leg to evade.

Both partners separate to reset in new ready position.

虎爪單刀對空手 二路
TIGER CLAW SABER VS. EMPTY HAND PART 2

Both partners complete rotation. **Saber** executes two reverse slashes, stepping into half horse with warlord bridge hand. **Empty Hand** executes tiger claw tearing strike, resetting into side stance with tiger claw bridge.

Saber jumps forward, executing slashing cut, immediately stepping up with right foot, chopping downward with vertical attack. **Empty Hand** shifts back and to right side, catching saber attack at wrist with left hand cutting down.

Empty Hand shifts body forward, attempting horizontal knife hand chop to neck of **Saber**. **Saber** ducks down to evade strike.

Immediately after ducking down, **Saber** slashes vertically upward. **Empty Hand** catches wrist of saber from underneath, flowing upward with slashing cut to deflect.

Empty Hand attempts to divert saber away towards his right.

Saber uses momentum to slash at left leg. **Empty Hand** raises leg to evade.

Empty Hand continues with momentum, spinning to right away from saber. **Saber** recoils slightly to his left, preparing for horizontal chopping strike.

Saber executes horizontal chop. **Empty Hand** blocks with double palm on forearm, strongly pushing away.

Saber uses pushing momentum to spin entire body around, executing horizontal chopping strike to head. **Empty Hand** immediately ducks down.

Following up immediately from high cut, **Saber** slashes at legs of **Empty Hand**. **Empty Hand** jumps high, drawing knees to chest to clear slashing cut.

Following up from slash to legs, **Saber** spins around to his right to execute horizontal cut to body. **Empty Hand** jams cut with right tiger claw heel kick to forearm.

Saber bounces back from heel kick to attempt second horizontal cut.

Empty Hand kicks forearm of **Saber** from underneath with left front kick to deflect.

Empty Hand charges forward with right front kick as **Saber** steps back with right foot into horse stance, cutting down with left knife hand to block.

Saber immediately after chopping foot rushes in to execute thrusting stab to torso. **Empty Hand** shifts body to his right in side stance, utilizing left tiger claw elbow at forearm to deflect thrust.

Saber immediately steps forward with left foot, chopping downward at right leg of **Empty Hand**.

虎爪單刀對空手 三路
Tiger Claw Saber vs. Empty Hand Part 3

Both partners jump and spin around to assume ready position, now on opposite sides.

Saber steps forward with left foot to execute high chopping strike to head as **Empty Hand** blocks with high X block.

Saber immediately follows up with left thrust punch to body as **Empty Hand** executes low X block.

Immediately, **Saber** executes left tiger claw roundhouse kick. **Empty Hand** shifts body, intercepting with outside tiger claw forearm deflection.

Without hesitation, **Saber** executes downward diagonal chop to head of **Empty Hand**. **Empty Hand** ducks down and away to evade.

Saber follows up with right tiger claw roundhouse kick. **Empty Hand** shifts to left, deflecting with double forearm block.

Saber immediately follows through with slashing strike to right leg of **Empty Hand**. **Empty Hand** evades by raising leg.

With no break in action, **Saber** immediately follows up with right diagonal cut. **Empty Hand** shifts and chops down at forearm of Saber with left knife hand cut.

Saber borrows energy of downward blocking knife hand for next attack. Both partners spin around, **Saber** to his left and **Empty Hand** to right.

Saber executes chop to waist from right side. **Empty Hand** shifts into left tiger stance, catching strike at wrist with hook hand.

Saber executes hook hand to remove **Empty Hand's** hook, preparing for next strike.

Saber draws back blade, preparing for next cut.

Saber executes horizontal cut to head of **Empty Hand**. **Empty Hand** ducks down under cut to evade.

Saber executes low chopping strike to right leg. **Empty Hand** raises leg, preparing to jump away.

虎爪單刀對空手 結論
Tiger Claw Saber vs. Empty Hand Closing Salutation

Empty Hand and **Saber** jump and spin back, returning to original positions. Both partners face each other in horse stance with tiger claw strike.

Both partners step backward to face each other in low back stance with high knife hand position.

Both partners rise up into left tiger stance, executing right and left downward facing claws.

Remaining in stance, both partners sink bridge hands.

Both partners draw hands inward to present tiger claw salutation.

103

單刀戰鬥技巧
Saber Techniques Against Staff

練刀之時。默想彼刀砍來。如何讓法。讓過之後。不容還手。緊逼一刀。所謂不招不架。只是一下。設若我刀為敵刀砍下。最好卽於砍下之處。隨時乘其虛隙以補進之。則較之雙方過門。再更他勢覷面接戰者。尤為巧妙。

When practicing the saber set, contemplate an opponent's attacks and your own defense. After defending against an attack, do not allow him to attack again, instead crowd him with your saber. As it is said: "If you do not spend your time defending, you will only need to deal with one attack. [Whereas if you are constantly defending, you will have to deal with many attacks.]" If an opponent slashes at me with his saber, the best thing to do is, just as his saber is about to reach me, to take advantage of the gap he has made and charge in through the doorway. By observing his attacks and engaging him in this way, you will become very skillful.
— 金一明 Jin Yiming 單戒刀 Single Defense Saber, 1932

The single saber, being one of the four essential basic weapons taught to all Kung Fu students regardless of style or system is an indispensable tool for the Kung Fu practitioner to be able to integrate mind, body and weaponry together as one. The saber can be pitted against empty hand or other weapons. In this section, we have chosen to display various techniques and applications pitted against the staff, which is another common weapon that all Kung Fu practitioners will need to learn and master. These two particular weapons are well suited to train against one another to establish a strong set of basic techniques and principles that will carry through into other areas of practice, be it empty hand or with weapon. Employed here are techniques displayed throughout the Black Tiger Single Saber set. Those striving to learn from this text can easily draw analogies from form to matching set to fighting application, thereby enriching their overall understanding and depth of the techniques of the single saber.

Image from 兵技指掌图说,
Illustrated Manual of Military Techniques by Naer Jing-e, 1843

The illusion of fighting, be it empty hand or weapon, as wonderfully choreographed in action movies, particularly of today with CGI graphics and other specialty tricks should be the furthest thing from our mind when we are talking about the true art of Kung Fu and particularly that of weapons fighting, speaking predominantly about the single saber. At no point should any true martial art practitioner be under the delusion that an actual combat situation will ever be like this. We must first take into account that we are not going to be fighting individuals that will be incompetent nor fail to fight back. When it comes to a combat situation, particularly on the battlefield where the matter is truly life or death, all fanciful and flowery movements will be put on the wayside. This is not to say that many of the actions found in weapon forms and saber sets are useless. The more fanciful movements are rather exaggerations of techniques that can be toned down and looked upon for their true core meaning rather than their dramatic counterpart that is put on display. The form is a training regimen which cannot be exactly equated with the way the weapon will be utilized in a combat situation. No one wants to be walking around the street hacking each other with sabers because the result is going to be a gruesome one, but this mentality must still be an ingredient in your process of thinking in order to maintain the martial quality of your practice. The line of demarcation must be made between performance and practicality when we talk about any type of fighting, and the practitioner must have a clear-cut understanding of this and be able to decipher what is what and its proper usage and timing. Again, this is not to say that particular movements that seem to be theatrical may not be applied, but it is definitely more dangerous and something to shy away from. The skill of the individual player most definitely will determine the outcome of a true combat situation.

Correct understanding and performance of the single saber must always be viewed with the enemy in mind. In this chapter, we have set forth several common scenarios of staff versus single saber to suggest some conceptual ideas of how particular movements of the single saber can be applied against various attacks from a staff. Both weapons can be considered either defender or attacker depending on the scenario. The applications displayed progressively become more complex. Some applications use a minimal amount of techniques while others will apply multiple techniques that can be either combined or used separately.

Keeping it simple is at the main heart of the issue when talking about combat with weaponry. Straightforward, direct and getting the task at hand accomplished is what we seek. In this way, every component of the single saber and the body of its practitioner will be in full use. There will be no aspect of this weapon not in play. All areas of the saber will be employed when executing an application with this weapon. Of course, the edge of the blade is prominent in the mind, but that is not to diminish the broad flat side of the knife, the handle, cup guard, tip and so on. There will be no situation where these varying areas of the knife will not be in play. The structure of the single saber lends itself to a wide variety of techniques that are effectively utilized. The myriad of possibilities and movements that can be done in such a scenario are countless and will vary greatly depending on timing, speed, distance, accuracy and competency of both combatants.

Practicality must be the underlying theme when reviewing either the empty hand matching set as presented earlier or these various applications against the staff. Although the saber fighter will have the advantage of the sharp blade in his hand, at no time should we become overconfident and feel that another combatant with a weapon will not be able to overcome our position. Therefore, refrain from flowery techniques that will destroy your ability to control the timing required to execute any practical movement. The blade's true nature is to cut. When fighting with the saber, we must press with our attack which also comprises our defense. With the combination of attack and defense in one action, the goal is to cut the opponent immediately and severely. This will include cutting at hands and legs as well as the more obvious cuts to the torso, waist, neck and head. This is the process of whittling down the opponent. A cut to the hand or leg may not necessarily finish the encounter but can do enough damage physically and mentally to break down their wherewithall to continue fighting, allowing the saber fighter to come in for the kill. Once we have been able to neutralize or disable the ability of the combatant against us, the saber practitioner can move in for the ultimate demise of the opponent.

Controlling the timing and space as well as the brevity of one's ability to engage with the staff attacker is at the heart of the issue. We must make sure that we close the gap between ourselves and them as soon as possible and get beyond the striking surface of the staff in order for us to be able to deflect, capture and employ our cuts as effectively and as soon as possible. Getting out of the range of the staff fighter or any other long weapon is important. We must be able to pass the two ends of the staff if being employed in double-head fashion and be able to bypass the single-headed position of either a long pole or spear-like halberd weapon in order to neutralize that threat. We must in all earnest attack any appendage that is closest to us to destroy further attack from the opponent. Unlike what some individuals may envision when they watch or practice their own single saber set, the distance between ourselves and the actual opponent is not that far. We will be fighting in an up close and personal manner. You will be able to see each other eye-to-eye, and this is the only way to inflict the damage necessary. Otherwise, we are purely executing a saber dance and exercise for our own benefit and not truly grasping the concepts and principles that are embedded within the saber form itself.

Displayed here are attacks from the staff in either a high, low, left or right scenario with the alternating sides of the staff or the actual tip itself thrusting out. Conversely, we also view the saber being the attacker and using the same four cardinal directional principles for attack and defense. The single saber will cover the three ranges of combat—long, middle and short. It is extremely useful in crossing over to the varying ranges for attack and defense.

Every facet of the body will be utilized. None will be spared. We will not only incorporate the saber itself, but we will use various hand and foot techniques from trapping, sweeping, locking, tripping and so on to get the task accomplished. The tiger claw techniques will be incorporated with every movement of the saber. Skillful footwork is key to the execution of the single saber in combat as it places oneself at close range and imminent threat. The agility and alignment of the saber practitioner is what puts the blade in the proper area to field and apply the technique. The saber will be an extension of the tiger claw movement and be incorporated fully, thereby allowing the entire body and mind of the tiger claw practitioner to execute each and every technique. We will not just rely upon one aspect of this weapon only, but incorporate our entire system into the movement.

The Black Tiger Claw Single Saber requires the practitioner to be ultimately alert. At no point can we be asleep or unobservant of what we are doing and the circumstances that we are in. In this way, practicing the empty hand versus saber set or these drills versus the staff will heighten the awareness of the practitioner to be better able to comprehend as well as understand the practice of the single saber, coming to the conclusion that the blade is as the practitioner should be—alive. This is the true concept behind the Black Tiger Claw Single Saber. It will adhere to the principles embedded in the Tiger Claw Kung Fu System and embody the ferocity of the attacking tiger that will be unrelenting.

黑虎爪單刀散手

Tiger Claw Saber
Applications

黑虎爪單刀-散手
SABER VS. STAFF APPLICATION 1

Both partners assume ready position.

Staff attacks, stepping forward with outside right strike. **Saber** shifts to left, intercepting with flat of blade.

Saber immediately grasps staff with left hand, executing chopping strike to neck.

黑虎爪單刀-散手
SABER VS. STAFF APPLICATION 2

Both partners assume ready position.

Staff steps forward with overhand strike to head. **Saber** blocks upward.

Saber grasps staff immediately with left hand, shifting forward, executing right diagonal chopping strike to neck.

黑虎爪單刀-散手
SABER VS. STAFF APPLICATION 3

Both partners assume ready position.

Staff steps up with right downward strike. **Saber** deflects upward with both left knife hand and saber to intercept.

Saber immediately grasps staff and executes low chopping strike to knee.

黑虎爪單刀-散手
Saber vs. Staff Application 4

Both partners assume ready position.

Staff steps up with low strike. **Saber** shifts forward into cross stance with low block to intercept.

Staff strikes high. **Saber** steps up into tiger stance with rising block to intercept.

Saber immediately grasps staff with left hand, executing chopping strike to leg.

黑虎爪單刀-散手
Saber vs. Staff Application 5

Both partners assume ready position.

Staff shifts into right bow, thrusting forward with strike. **Saber** shifts back into horse, deflecting to left with flat of blade.

Saber grabs staff, immediately executing stabbing strike to throat.

黑虎爪單刀-散手
Saber vs. Staff Application 6

Both partners assume ready position.

Staff thrusts forward with strike. **Saber** shifts to left, intercepting with base of blade.

After intercepting staff strike, **Saber** rotates entire body inward toward **Staff**.

Completing rotation, **Saber** wraps left arm around staff, simultaneously executing chopping strike to neck.

黑虎爪單刀-散手
Saber vs. Staff Application 7

Both partners assume ready position.

Staff steps forward with downward strike. **Saber** intercepts with rising knife hand and saber.

Staff immediately shifts body backward, executing low sweeping strike to leg of **Saber**. **Saber** evades sweep.

Saber immediately follows up with downward chopping strike to base of neck.

黑虎爪單刀-散手
Saber vs. Staff Application 8

Both partners assume ready position.

Staff steps up with right strike to head. Saber deflects with left arm and saber simultaneously.

Immediately, Saber attempts downward chopping strike. Staff steps back, blocking with left side of staff.

Saber uses left hand to shear off staff, immediately grasping it and wrapping staff under arm to execute downward chop to neck.

Maintaining control of staff, **Saber** follows up with upward slashing cut to neck.

黑虎爪單刀-散手
Saber vs. Staff Application 9

Both partners assume ready position.

Staff shifts forward, executing high stabbing strike. **Saber** shifts to left, deflecting with base of blade.

Staff immediately executes second thrusting strike. **Saber** deflects to right with base of blade.

Saber immediately grasps staff from left side, trapping underneath, then executes chopping strike to neck.

Maintaining control of staff, **Saber** turns around, slashing at back of leg.

Saber rotates body around to capture **Staff** at neck and executes reverse sweep with left leg, throwing **Staff** to ground.

Saber follows up with downward chop.

黑虎爪單刀-散手
Saber vs. Staff Application 10

Both partners assume ready position.

Staff steps forward, executing downward strike to head. **Saber** shifts body into reverse cross stance, intercepting with left hand and blade together to redirect staff.

Immediately, **Saber** slides left foot out and turns blade upward in reverse position to cut throat.

Following up, **Saber** executes right cross strike with pummel of blade.

Reversing direction, **Saber** applies slashing strike to right side of neck, pressing blade with left hand, drawing backward.

Finishing up, **Saber** raises blade and executes final chopping strike to base of neck.

黑虎爪單刀-散手
Saber vs. Staff Application 11

Both partners assume ready position.

Staff shifts forward, thrusting strike to leg. **Saber** side steps with tiger stance, deflecting outward with dull side of blade.

Staff shoots thrusting strike upward. **Saber** shifts again, deflecting high with flat of blade.

Saber immediately uses left hand to capture staff.

Controlling staff, **Saber** executes right and left slashing strikes to waist.

Maintaining control of staff, **Saber** spins around, slashing at leg.

Saber follows up with thrusting strike to throat.

黑虎爪單刀-散手
Saber vs. Staff Application 12

Both partners assume ready position.

Staff thrusts strike to torso. **Saber** shifts backward, deflecting with base of blade.

Saber immediately slides left foot backward and begins spinning to rotate round, executing slashing cut to head. **Staff** ducks down.

Staff rises up, executing downward left hand strike. **Saber** utilizes left knife hand and blade to intercept.

Saber immediately wraps left arm around staff to control and steps forward, executing downward chopping strike to neck.

Immediately following up, **Saber** executes right cross strike with pummel to jaw.

Saber, maintaining control of staff, swiftly executes right foot sweep to unbalance **Staff**.

Disarming **Staff**, **Saber** immediately follows up with downward chop to neck.

黑虎爪單刀-散手
Saber vs. Staff Application 13

Both partners assume ready position.

Staff executes thrusting strike. **Saber** steps to left, deflecting outward with base of blade.

Swiftly, **Staff** executes second thrust. **Saber** shifts body and deflects inward with base of blade.

Saber steps forward, utilizing left hand to open staff and executes horizontal cut to head. **Staff** ducks to evade.

Saber immediately follows up with inside slash to leg. **Staff** raises leg to evade.

Saber drops down to kneeling position, immediately thrusting forward to body with stabbing strike.

黑虎爪單刀-散手
Saber vs. Staff Application 14

Both partners assume ready position.

Staff executes low stab. **Saber** moves in, rotating saber downward to deflect with dull back of blade.

Without hesitation, **Saber** leaps forward, executing circular downward strike to cut hand of **Staff**.

Saber follows through with downward stab to throat.

黑虎爪單刀-散手
Saber vs. Staff Application 15

Both partners assume ready position.

Staff thrusts out with low strike. **Saber** deflects to left with dull of blade.

Staff executes second thrust. **Saber** shifts back and deflects to other side with base of blade.

Staff executes third thrust strike to body. **Saber** raises up left hand and blade to deflect. Immediately, saber captures staff and executes downward cut to neck.

Disarming **Staff**, **Saber** swiftly executes right heel kick to body.

Saber immediately spins around, executing left tiger tail kick.

黑虎爪單刀-散手
Saber vs. Staff Application 16

Both partners assume ready position.

Saber steps forward, executing thrusting strike. **Staff** shifts to left and blocks thrust.

Saber steps up with left and executes second thrust. **Staff** shifts to right, deflecting.

Staff steps up, executing downward strike. **Saber** shifts backward, deflecting with base of blade.

Saber uses left hand to capture and drive staff down.

Maintaining control of staff, **Saber** spins body around to execute right foot sweep.

Saber follows up with undercut slash to neck.

Saber finishes with downward chop.

黑虎爪單刀-散手
Saber vs. Staff Application 17

Both partners assume ready position.

Saber leaps forward, executing downward chopping strike to head. **Staff** blocks upward.

Saber steps forward, executing low stabbing thrust. **Staff** steps back into cross stance, blocking low.

Staff continues and sweeps away saber thrust.

Staff continues motion, spinning around, executing left hand thrust. **Saber** shifts to right, deflecting with flat of blade.

Saber spins around, slashing at leg.

Saber continues rotation, capturing **Staff** from behind, slitting throat.

黑虎爪單刀-散手
Saber vs. Staff Application 18

Both partners assume ready position.

Staff steps up with same foot, executing right side strike. **Saber** shifts to left, deflecting with flat of blade.

Staff executes low strike to leg. **Saber** withdraws stance and deflects outward.

Staff executes third strike directly down. **Saber** captures with left hand.

Maintaining control of staff, **Saber** executes diagonal chop to neck.

Following up immediately, **Saber** executes downward stomping kick to knee.

Spinning around, **Saber** disarms **Staff** and executes final downward chopping strike.

黑虎爪單刀-散手
Saber vs. Staff Application 19

Both partners assume ready position.

Saber charges forward with wide arching overhand chop. **Staff** shifts to right, knocking saber to ground.

Saber recoils and attempts second wide arching downward chop. **Staff** shifts to left, knocking saber to ground.

Staff immediately rushes forward to bar **Saber's** possible attack. **Saber** uses both hands and blade to jam.

Immediately, **Saber** uses cup guard to hook away and disarm **Staff**, striking with pummel.

Saber spins around, executing spinning hook kick. **Staff** ducks down to evade.

Saber immediately follows through with left foot sweep, taking **Staff** to ground, spinning around, executing final chopping strike.

Conclusion

練刀一道。在昔專為禦敵而用。故不得不如上述所云。欲心領神會。揣測來去之方與砍刺之法。然近世射擊之學盛行。鎗彈之效尤著。練刀一事。不過增長體力膽力。與靈活身腰。流通血液。養成民眾有果敢冒險之心而已。若持四尺之單刀與三寸之手鎗以相抗衡。則螳臂當車。未免不自量矣。

The way of saber training was previously for the specific use of defending against enemies. Therefore everything described above was a matter of course, understood intuitively – observing and anticipating an opponent's directions of withdraw and attack, the methods of slashing and stabbing, etc. However in modern times, training in firing guns is more popular, bullets being so effective. The purpose of practicing the saber art is only to increase strength and courage, to make the body more nimble, improve circulation, and to cultivate in ordinary people the daring to face challenges. If you were to take up a four-foot saber and contend against someone with a three-inch pistol, you would be overestimating your ability. Avoid doing that.

— 金一明 Jin Yiming 單戒刀 Single Defense Saber, 1932

We can now see in this modern day and age that the traditional art of Kung Fu weapons and their practice still holds a strong and meaningful place in our day-to-day training. Although they may not be needed as their original purpose may have been deemed, the training of weaponry benefits the traditional martial arts practitioner in countless ways not obvious at the start. The fusion of mind, body and saber give the Kung Fu student the chance to go beyond themselves in thought and expression, while the training also reinforces all the necessary qualities to further one's deep understanding of not only the execution of the single saber and its way but also to delve deeper into the true spirit and movement of the tiger that resound throughout.

The practitioner has to spending great time and effort in researching, practicing and studying the weaponry within the art of Kung Fu and particularly the single saber as it is related to the techniques and actions of the Tiger Claw System. I have come to the conclusion that in the hands of a well-trained and knowledgeable practitioner, the single saber must wholeheartedly and without question be alive in all senses of the word. There will be no facet of the practitioner and his saber that will not be intertwined, connected and feeding off one another. The saber practitioner and his weapon are succinctly conjoined and will flow with every motion uninhibited. We see now that through a thorough training with the principles of the Tiger Claw System, the saber set is a unique extension of each and every movement utilized in this dynamic art form. There will be no gap between the player of the Tiger Claw System and the saber. Each one is incorporated into the other.

This understanding and feeling of being connected to one's weapon or tool is what makes the study and practice of Kung Fu weaponry so unique and fosters all other aspects of our martial training, not unlike the tools of a craftsman or world class cyclist, race car driver and so on. The skill of the Kung Fu practitioner and his connection to his weapon must be indelible to the point that we cannot tell where the movement is emanating from and resides.

When we speak about the concept of being alive, in touch or in tune with one's tool or weapon, it is the most difficult thing to explain to the uninitiated individual that has never striven to attain this type of connection. It goes far beyond the mean and meager execution of thrashing about with the knife, but rather each and every cut and stroke is neither over nor underdone and is a true expression not only of the saber player's mind but a reaction and reflection to what is necessary to be done and executed, understanding that when they play their set, the invisible opponent is ever-present in their mind.

We can truly say that the saber is an extension of the mind and body of the practitioner and vice versa where the saber is the inspiration and sets a light in the mind of the individual. Once the unification of mind, body and weapon is achieved, we can begin to understand how to embrace the ferocity and unrelenting will of the attacking tiger that is inherent within the movement of the saber, thereby adhering to not only the concept of the tiger's attack, but bringing it forward in full exhibition with the execution of the Black Tiger Claw Single Saber and its standing within the Dynamic Art of the Tiger.

About the Author

Master Paul Koh is internationally recognized as one of the world's leading Kung Fu teachers. He has dedicated his life to the study of this ancient Chinese art form, immersing himself in Chinese language and culture since his early teens. He is one of the only major advocates for preserving and promoting this cultural art which, despite being frequently overshadowed today by popular fads, is a timeless and ever-evolving art that goes far beyond a punch and kick.

With over 40 years of experience, Master Koh has extensively trained with world-renowned Kung Fu masters in 虎爪派 Fu Jow Pai Tiger Claw Kung Fu, 洪家拳 Hung Gar Tiger Crane. He has expertise in classical Chinese weaponry and in traditional southern lion dancing. Master Koh is not only a martial artist who has dedicated countless hours, days, months and years to the study of his discipline, he is also a true educator, mentor and motivator. He has taught thousands of students and currently maintains a martial art training hall in NYC Chinatown. His knowledge of Kung Fu goes far beyond the physical execution or application of a technique, and extends to the historical and cultural background of Chinese martial arts, as well as the deep philosophical bedrock upon which this art form depends. These qualities, combined with an ability to speak and write clearly and eloquently, make Master Koh a unique phenomenon in the martial arts world.

Practicing Tiger Claw Saber in Grandmaster Tak Wah Eng's studio

In addition to studying, researching and teaching the Chinese martial arts for nearly four decades, Master Koh is an accomplished author, having written several previous texts on the art of Kung Fu, as well as having published many articles in various martial art magazines throughout the years. In 2018, he launched Kung Fu In A Minute, a unique approach to documenting, promoting and preserving the traditional skills, attitudes, traditions and practice of the art of Kung Fu. Through Kung Fu In A Minute, he is creating a library of videos, articles and manuals to further promote, preserve and protect this ancient art form.

Master Koh demonstrating Tiger Claw Saber at New York Asia Society, Chinese New Year 2018

Additional titles available from

Kung Fu In A Minute Publications and Master Koh

十獨手
*Dynamic Art of the Tiger:
Ten Essential Techniques of the Tiger*

虎爪木人樁大師
*Dynamic Art of the Tiger:
Tiger Claw Wooden Master*

八卦虎爪對拆
*Dynamic Art of the Tiger:
Tiger Claw Eight Diagram Fighting Set*

虎爪八卦棍
*Dynamic Art of the Tiger:
Tiger Claw Eight Diagram Long Pole*

嶺南拳術五行拳
*Southern Shaolin:
Five Element Fist*

嶺南拳術仙鶴拳
*Southern Shaolin:
Immortal Crane Fist*

嶺南拳術虎爪手法
*Southern Shaolin Tiger Claw:
Principles of the Tiger*

猛虎鐵鎚
Fierce Tiger Iron Hammer

嶺南虎爪雙刀
*Southern Shaolin:
Tiger Claw Double Knives*

虎鶴對拆
*Southern Shaolin:
Tiger & Crane Matching Set*

For more information, visit
KUNGFUINAMINUTE.COM